A baby bison stays close to mom during a May snowstorm near Black Sand Basin. Spring comes late in Yellowstone.

Left: A rainbow forms while Castle Geyser erupts in the Upper Geyser Basin. Castle is easy to predict, with eruptions every 10-12 hours.

VISIONS OF YELLOWSTONE

A JOURNEY THROUGH THE WORLD'S FIRST NATIONAL PARK

PHOTOGRAPHY BY TODD ANDERSON

HARMONY HOUSE PUBLISHERS

DEDICATION

For Mom, Dad, and Sister

Harmony House Publishers

P.O. Box 90, Prospect, Kentucky 40059

(502) 228-2010 or 228-4446

©2003 Todd Anderson

Photography ©2003 Todd Anderson

Library of Congress 2003102792

ISBN 1-56469-105-5

Executive Editor: William Strode

Design: Karen Boone

Printed in Hong Kong

*Hikers near the summit
of Avalanche Peak (right).
The 10,566-foot high peak has
great views. Grizzly bears
are occasionally spotted in
this area.*

*Page One: A bull elk bugles
loudly near Norris Campground.*

*Title Page: The Lower Falls of
the Yellowstone River as seen
from Artist Point. This is literally
where you can see yellow stone.*

Contents

Fishing Cone steams at sunrise in the West Thumb Geyser Basin.

A grizzly bear (right) makes it's way through spring snow near Obsidian Cliff. Seeing a grizzly bear in the wild is an incredible experience.

Every day a new picture is painted and framed,
held up for half an hour, in such lights as the
Great Artist chooses, and then withdrawn, and
the curtain falls. And then the sun goes down,
and long the afterglow gives light.

—Henry David Thoreau

In this lengthy time exposure
steam from Old Faithful Geyser
drifts in the air as the lights of
Old Faithful Inn come on.

The sun was beginning to set as I drove the curving roads of Yellowstone National Park to Old Faithful, feeling very tired but also very good. I had finally photographed the elusive grizzly bear for three straight days now, a goal of mine for years. I felt so alive at that moment. It's spring and grizzlies were close to the road. One of the foremost experts on grizzlies told me go to Alaska because it's just too hard to get good photos of them in Yellowstone,and I had begun to believe he was right and then.... • It was still dark out in Madison Campground and the temperature was 20 degrees on this spring morning. I crawled out of my warm sleeping bag and raced to throw my tent in the car. I had information that a grizzly sow with two cubs was near the road by Swan Lake but I needed to get a very early start to obtain beautiful early morning light. • After driving for about 50 minutes I pulled off the road where I think the bears might be. Thick gray clouds cover nearby mountains—so much for great light. However, something brown moves in the sagebrush. Time to break out the 600mm lens with 1.4 extender. I focus, look closely for the hump. It's there, and this is definitely a grizzly bear! • These first photos probably won't work as it's too dark out. Patience. A wolf comes very close to the bear, and the light gets better for photos.

Hours pass by like minutes as the wolf and grizzly interact. A large group of photographers, tourists, and rangers look on. Over the next three days more grizzly bears will come close to the road for some wonderful photos. • I was ten years old when I took my first trip to Yellowstone. That summer my family loaded up the station wagon in Indiana and pulled a camper behind it. Since then I have made many trips and lived close to the world's first national park. Yellowstone holds a very special place in my heart. • I have divided this book into four geographical sections of Yellowstone. In each section I describe my favorite places and show you photos of those areas. *The Greater Yellowstone Ecosystem*, which includes the Tetons is a very large area so I am just giving you my favorites. I'm sure you will see and photograph many other places of Yellowstone's extraordinary beauty. • My hope is that after seeing these photographs you will be inspired and, like me, want to protect this incredible place for future generations.

—Todd Anderson

Winter time magic: A rainbow is forming above the road near Madison.

Fittingly we start our journey through Yellowstone at the Old Faithful area. Old Faithful Geyser, uncontrolled by man, goes off 365 days a year, day and night. To leave the crowds behind, try watching the world's most famous geyser at sunset, sunrise, or on a full moon night. There are many other geysers to see from the boardwalks so be sure to stop in the Old Faithful Visitor Center where you can jot down the predicted eruption times. Beehive, Castle, Grand, and Riverside Geysers are favorites and these can either be seen in a short walk or combine them with Biscuit Basin for a half day hike or cross-country ski. For a special treat get the kids of all ages some ice cream at the general store or Old Faithful Inn after geyser gazing from the boardwalks. • Work began on Old Faithful Inn in 1903 and went on through the very cold winter and opened for guests in 1904. The Inn is made of logs cut from nearby trees and local stone was used for the massive fireplace. It blends into its surroundings beautifully and is one of my favorite places to stay. • When it's time to head north from Old Faithful, you will pass the beautiful Black Sand and Biscuit Basins. If you didn't walk to them from the Old Faithful Inn, please stop off. Next comes Midway Geyser Basin and its incredibly colorful Grand Prismatic Springs. Firehole Lake drive is next, with the Great Fountain Geyser and it's reflecting pools of water. Thermal features are a very unique part of this section. Bison and elk are commonly seen close to the geysers. • Pretty Madison Campground sits next to the Madison River and is the closest campground to Old Faithful. Going left at Madison will take you 14 miles to the fun town of West Yellowstone. From Madison take the road 14 miles to Norris, passing through the beautiful Gibbon Canyon and make a stop at Gibbon Falls. • Norris Geyser Basin is well worth a walk around it's boardwalks. Echinus Geyser and the great view from the museum overlook of Porcelain Basin are favorite scenes of mine. A campground is located across the road from Norris Geyser Basin. The Gibbon River and it's meadow are next to the campground and elk or bison wander through in the evenings. •

Old Faithful Geyser at sunset. (left)
Members of the 1870 Washburn Expedition
named the geyser for its consistent eruptions.
Bikers (above) watch Old Faithful erupt.

A full moon rises as
geyser steam drifts
into the air in the
Upper Geyser Basin.
Sunset or sunrise is
a great time to stroll
the boardwalks here.

Alpenglow on the
Firehole River (right) in
the Upper Geyser Basin.

A young girl watches
as *Castle Geyser* erupts in
the Upper Geyser Basin.
Brilliant blue sky (right)
emphasizes white spray
from *Castle Geyser*.

A bison (left) grazes in front of the Old Faithful Inn. Bison and elk commonly come very close to the Inn. The unusual clock (far left) is in the lobby of the Old Faithful Inn.

The lobby of the Old Faithful Inn (right) is seven stories tall and has a gigantic fireplace. This is one of my favorite buildings, and I love to sit in the lobby and people watch.

During the long winters inn employees have a little fun. Here a worker blows snow towards a co-worker from the front porch of Old Faithful Inn. Employees taking down flags have a great view (right) from the roof of Old Faithful Inn.

Much of the Old faithful Inn (left) was built in the winter of 1903-1904 using local timber and stone.

A couple watches Old Faithful erupt. Another great view (right) of Old Faithful from the porch of Old Faithful Inn. The porch offers good views of the Upper Geyser Basin too.

The sun sets behind Great
Fountain Geyser in the
Lower Geyser Basin.

Colorful Grand Prismatic
Spring in the Midway
Geyser Basin (left) is one
of the world's largest .

Muddy water splashes out of Yellow Funnel Spring in the Norris Geyser Basin.

Hot water (right) splashes in the Black Warrior Geyser Group on Firehole Lake Drive.

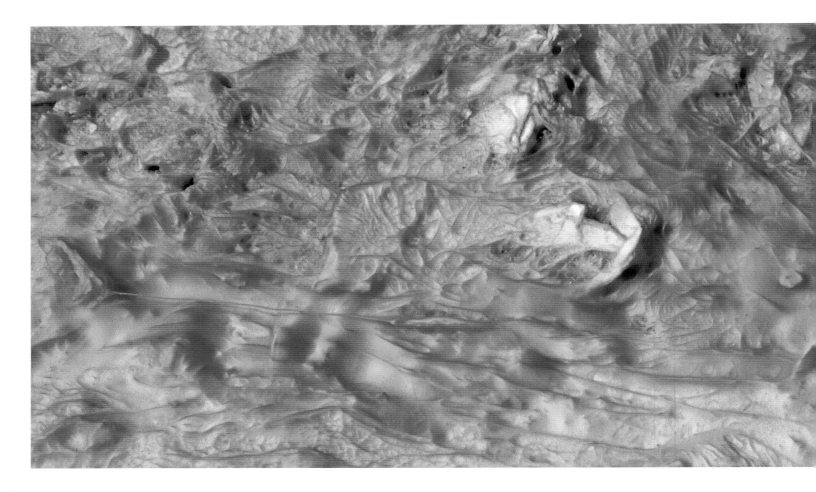

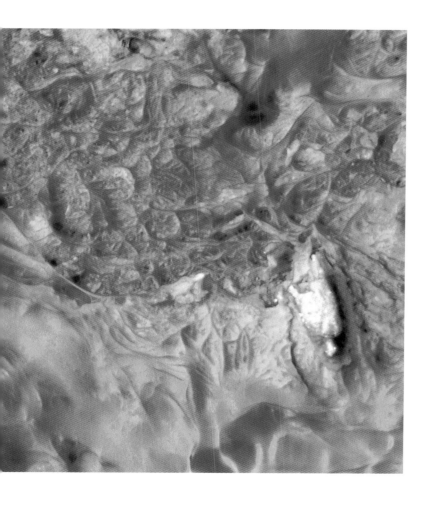

How deeply with beauty is beauty overlaid!

—John Muir

A close-up of Grand Prismatic Spring. Thermophilic cyanobacteria (microscopic organisms) help create the orange color.

A Yellowstone National Park Ranger takes scientific readings, (such as the temperature of the water), in the Norris Geyser Basin.

Solitary Geyser erupts (left). This geyser sits off by itself in the Upper Geyser Basin and is well worth the short walk to see it.

Elk cross the Firehole River as Castle Geyser erupts in the background. Bison and elk are commonly seen in the Upper Geyser Basin year round. Are they as impressed as humans when the geysers blow?

Elk (right) beside the road near the Madison Campground.

I think I could turn and live with animals,
they are so placid and self-contain'd;
I stand and look at them long and long.

—Walt Whitman

*Bison gather as large plumes
of steam rise from the Lower
Geyser Basin.*

A bison near Fountain Flat Drive. Clouds
float behind a bison (above). Bulls can
weigh up to a ton and are considered the
largest land mammal in North America.

A child (far left) takes in one of the
many wonders of Yellowstone
on the Fountain Flat Drive.

Interesting patterns of water and steam reflect the sunset in Norris Geyser Basin. Cooling off on a hot summer day (right), a young woman soaks her feet in the Firehole River off the Firehole Canyon Drive.

As Riverside Geyser goes off, a rainbow forms. This geyser is next to the Firehole River in the Upper Geyser Basin.

Grand Geyser (right) erupts in full glory. I like to go to the Old Faithful Visitor Center and write down the predicted times for the major geysers. But on this evening, luck was on my side, because I had no idea when Grand would be erupting.

The sun rises behind Castle
Geyser in the Upper Geyser Basin.
The geyser was named for it's
resemblance to a ruined tower
of an old castle.

A well prepared observer
(right) stays dry under an umbrella
as Beehive Geyser erupts over
the boardwalk in the Upper Geyser
Basin. Because Beehive blasts
water so high into the air, the
water cools as it falls toward
people on the boardwalk. If I had
to pick my favorite geyser it
would be Beehive.

Winter lingers into May
as bison graze near the
Firehole Lake Drive.

Part of a lone tree (left)
pokes through winter
snow at the Fountain
Paint Pot area.

In wildness is the preservation
of the world.

—Henry David Thoreau

Coyotes make their way
through winter snow near
Madison. Wildlife are easier
to see and photograph
in the winter.

A frosty bison lies near Old Faithful Geyser. Cold winter mornings can be surreal near geysers, as thermal steam rises and freezes creating a rime-laden fog.

The view in front of a snowcoach as bison pass by on the road from Old Faithful to Madison. The bison use the plowed roads to travel easier in winter.

A ranger (above) on cross-country skis observes Riverside Geyser erupt. A cross-country skier (right) looks at trees coated by a thick white frost in the Upper Geyser Basin.

Snowmobilers (far right) leave their vehicles to watch bison cross a boardwalk in Biscuit Basin.

YELLOWSTONE LAKE TO GRAND TETON

This section offers a mix of road travel and beautiful hiking and backpacking. Coming from the east entrance the road climbs steeply over Sylvan Pass and leads to Avalanche Peak, a favorite hike. Park just west of Eleanor Lake to start this strenuous half day hike with an elevation gain of 2,150 feet. You will find some incredible views of the Absaroka Mountain Range and Yellowstone Lake. Keep an eye out for grizzly bears. I also like to bring my lunch. The summit of Avalanche Peak is a great place to dine. • Continuing on, you soon reach massive Yellowstone Lake. Bison like to wander close to the road near Sedge Bay. Next is Pelican Valley on the north side of Yellowstone Lake, home to many hungry grizzly bears. At Fishing Bridge, turn south. Great places to stay are the Lake Hotel or Bridge Bay Campground. The vistas from the hotel or campground of Yellowstone Lake are awe-inspiring, especially at sunset. The road hugs the shoreline of Yellowstone Lake for the next 21 miles to West Thumb Geyser Basin. This geyser basin is worth a stroll for great views of Yellowstone Lake. Close by is Grant Village where good camping sites are available. • Heading west, it's 17 miles to Old Faithful as you cross the Continental Divide twice. I parked at the DeLacy Creek trailhead for a three day backpack trip into Shoshone Lake and the Shoshone Geyser Basin. After 7.5 miles I stopped for the first night and camped along the shore of the lake near it's narrowest point. Backcountry permits for this trip can be picked up at either the Old Faithful or Grant Village Ranger Stations.

The next day was a short 3-mile hike to my campsite close to the shore of the lake. This left plenty of time to hike about one mile and explore the Shoshone Geyser Basin. Kayaking and canoeing are popular on Shoshone Lake, I saw a couple kayaking to the basin the next morning. On my third day 8.5 miles of pleasant hiking lead me to finish at the Lone Star Geyser trailhead. A ride in the back of a pickup truck from a friendly couple from Idaho got me back to my car. • There is a parking area 7 miles south from West Thumb. This is the start of another favorite three day backpack trip to Heart Lake. On the second day it's a steep climb up to the Mt. Sheridan fire lookout for great views of the Tetons and Yellowstone's lakes. Heart Lake Geyser Basin is close to the campsites and fun to explore. It's 8 miles to the campsites and another 3 miles to the top of Mt. Sheridan. • The road goes by beautiful Lewis Lake. The south entrance is 22 miles from West Thumb. It's only a short 8 mile drive to Grand Teton National Park from the south entrance of Yellowstone. Many people go to both parks in the same trip because they are so close together. The views of the Tetons are spectacular as you drive along Jackson Lake and Jenny Lake. •

Layers of fog float over Heart Lake (left) as the sun rises on
Mt. Sheridan. Steam rises from a geyser on right side of picture.
A backpacker (above) takes in the view at Shoshone Lake.

Splendor of ended day, floating and filling me!...You, Earth and Life, till the last ray gleams, I sing.

—Walt Whitman

A lone duck floats by as the moon rises over Heart Lake in Yellowstone's backcountry.

A close-up section (left) of a tree next to Shoshone Lake shows interesting lines.

Sunset paints the sky with pastels near Bridge Bay Campground. A fitting goodbye, this picture was taken on the last evening of my last trip to Yellowstone.

Boiling water and steam spew near Shoshone Geyser Basin as kayaker's approach.

Room with a view: A backpacker (left) takes in Shoshone Lake.

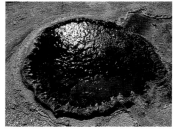

Water boils in a hot spring (above) as Minute Man Geyser (left) erupts. Both are in the Shoshone Geyser Basin. Please be careful. There are no boardwalks in the basin.

Colorful runoff (right) channels flow towards Yellowstone Lake in the West Thumb Geyser Basin.

On land only the grass and trees wave,
but the water itself is rippled by the wind.
I see the breeze dash over it in streaks
and flakes of light.

—Henry David Thoreau

A different kind of traffic jam forms as bison use the road on the north side of Yellowstone Lake.

A bicyclist at Bridge Bay Campground.

Fishermen (left) at Yellowstone Lake near Grant Village.

The classic view of the
Teton Mountain Range from
Snake River Overlook made
famous by Ansel Adams
(photograph).

Alpenglow fills the sky
above Jackson Lake (left)
in this view from Signal
Mountain in Grand Teton
National Park. Only 8 miles
separate the two national
parks—Yellowstone and
Grand Teton—so many
people visit both on the
same trip.

God beams break
through the clouds near
Mt. Moran in Grand
Teton National Park.

A flock of birds (right)
fly in the last glow of
sunlight near Oxbow
Bend in the Tetons.

Otters swim in Jackson Lake (left).
A swimmer leaps into String Lake
in Grand Teton National Park.

The Tetons (left) rise high above
the sagebrush near Mountain
View Turnout.

Wherever we go in the mountains
we find more than we seek.

—John Muir

A moose looks for food at
twilight near Jackson Lake
Lodge in the Tetons.

The Hayden Valley to Lamar Valley section is a good place to spot wildlife. Going north from Fishing Bridge, it's 16 miles to Canyon Village. The road closely follows the Yellowstone River where fishermen and white pelicans are frequently seen. Hayden Valley starts just after the Mud Volcano. Large groups of bison roam this valley and often stop traffic as they cross the road. Also keep a lookout for moose, elk, coyotes and grizzly bears. Make sure to bring binoculars for extra close-up views of Yellowstone's wildlife. • The best times to see wildlife here or anywhere in the park are early mornings and evenings until sunset. Please be polite and pull off the road at parking areas when you see wildlife. • Shortly after leaving the Hayden Valley is a turnoff for South Rim Drive. This short drive will take you to Artist Point, my favorite place to observe the Grand Canyon of the Yellowstone and Lower Falls. Here you really can see the "yellow stone" caused by hot water acting on volcanic rock. To leave the crowds behind take either the North or South Rim Trails. The trails start from either Canyon Village or Artist Point. The trail near Artist Point is where Thomas Moran made sketches and took notes from which he would later paint his famous work of Lower Falls and the Grand Canyon of the Yellowstone. • Get ready for switchbacks as you leave the canyon and travel the 19 miles over steep Dunraven Pass.

My favorite half day hike in Yellowstone starts at a parking area at the summit of Dunraven Pass. It's 3 miles up an old roadbed to the summit of Mt. Washburn at 10,243 feet. Bighorn sheep are sometimes seen near the summit. The vista from the fire lookout station is awe-inspiring in every direction. • Tower Falls is up next and I recommend the short hike to the base of the falls where a rainbow is sometimes seen. Good camping is also available at Tower Falls. From Tower/Roosevelt it's 29 miles to the Northeast Entrance. • The road soon follows the Lamar River and the Lamar Valley comes into view. The valley is a great place to spot Yellowstone's wolves and grizzlies. Again, remember to bring binoculars and use them in the early morning or near sunset. Antelope, bison, coyotes, and fishermen also love this valley. • The Yellowstone Association Institute which offers classes on just about anything to do with Yellowstone (animals, geology, photography, etc.) is located at a spot along the road where wolves are frequently seen. The tall and beautiful Absaroka Mountains come into view as you leave Lamar Valley. •

In the Lamar Valley, (left) the sun sets
behind a colorful display of summer wildflowers.
Fly Fishing (above) in the Lamar River.

Storm clouds cover
a mountaintop behind
brilliant fall color in
the Lamar Valley.

Bison graze (right)
in the Lamar Valley
at sunset.

...All my life I've been growing
fonder and fonder of wild places
and wild creatures.

—John Muir

A lone bison wanders
near the Lamar River.

Wildlife watchers in the Lamar Valley at sunset. Pronghorns (left) are a common sight in the Lamar Valley.

Winter and summer
transportation in the Hayden
Valley: A snowcoach (top)
uses the same road as cars
and buses as they wait on
a bison crossing.

Bison use their large heads
(left) to push away deep winter
snow as they search for food
in the Hayden Valley.

The 10,243-foot summit
of Mt. Washburn,
has great views in all
directions. Hikers (right)
descend from the Mt.
Washburn summit as
storm clouds gather in
the distance.

Bison also known as buffalo run through fresh snow during a spring snowstorm.

A mule deer (right) moves quickly through the snow in Yellowstone's high country.

From Norris, go east for 12 miles to Canyon. In this short drive keep an eye out for elk in the meadows along the Gibbon River and take the one way road to Virginia Cascade. The other option from Norris is to head north 21 miles to Mammoth Hot Springs. I covered the Norris Geyser Basin in a previous chapter. • Shortly after passing Roaring Mountain you come to Obsidian Cliff. Obsidian is a volcanic glass that Native Americans used for projectile points and cutting tools. Not too far from Obsidian Cliff is appropriately named Grizzly Lake. This section of the park, (up to Mammoth Hot Springs) is an area where I saw three grizzlies. Remember, no close-up picture is worth risking your life. Grizzlies can run faster than man. Although they rarely attack, grizzlies have injured and killed people in Yellowstone. • You will pass Willow Park a good place to look for moose. Sheepeater Cliff has picnic tables for lunch or dinner. Keep an eye out for wolves as you approach Swan Lake. Beautiful Electric Peak draws your eye to the northwest. • The park headquarters are located at Mammoth Hot Springs. Large groups of elk are usually munching on grass next to the buildings. The hot spring terraces are

worth a stroll on the boardwalks. A good campground is located here along with a hotel. If the kids are hungry stop in either the general store or the grill for ice cream. There's something very special about eating ice cream while watching elk. But don't feed the elk ice cream! The Albright Visitor Center has wonderful exhibits about the history of the park. • From Mammoth it's 5 miles to the north entrance and the town of Gardiner. Heading east out of Mammoth it's 18 miles to Roosevelt. Keep an eye out for bison in this area. The one-way Blacktail Plateau Drive, a dirt road, is especially pretty in the fall. Look for mule deer, elk, and pronghorns. • There are many good books to guide you on more specific aspects of Yellowstone. Two of my favorites are: *Day Hiking Yellowstone* by Tom Carter and the *The Geysers of Yellowstone* by T. Scott Bryan.

The last of fall color at Blacktail Deer Plateau.
Steam rises from Mammoth Hot Springs (left)
as an elk stands nearby.

A cold fall night coats
trees with frost near Indian
Creek Campground.

An autumn sunrise
(left) highlights aspens
near Roosevelt.

The claws of a grizzly bear hold part of his lunch. Claws make for a good grip in the snow as a grizzly (right) makes it's way across a slope in the northwest part of Yellowstone.

A young elk at dawn near Beaver Lake.

The new antlers in velvet—an elk (left) during a spring snow near Blacktail Deer Creek.

In this sequence of photographs a grizzly bear feeds on a carcass near Swan Lake while a wolf also tries to feed. I spent most of a thrilling day watching this game of cat and mouse, as the wolf was occasionally allowed to eat.

Hot water flows into the Gardner River's cool water: A terrific spot to watch a sunset.

A father gives his son a lift (right) while fishing near the Indian Creek Campground.

A baby bison jumps near the Gibbon
River. An elk grazes (above) in front
of a church at Mammoth Hot Springs.

On a cold fall evening (far left) a camper
cooks dinner at the Mammoth Campground.

If a war of races should occur between the wild beasts and Lord man, I would be tempted to sympathize with the bears.

—John Muir

Two young grizzlies play in the snow near Swan Lake after their mother left them on their own. The day before I watched the bears and their mom cross a snowy mountain slope too far away for pictures.

Fresh snow covers the
nose of a grizzly bear.

During a spring sunset (right)
a lone grizzly crosses a field
of snow in the northwest
part of Yellowstone.

Grizzly bears, like humans, can do funny things, as this bear demonstrates in the northwest part of Yellowstone.

A black bear (left) with her two cinnamon colored cubs look for food at the Tower-Roosevelt area.

I used a wide variety of camera gear for the photographs in this book. 35mm cameras used were the Nikon F100, Nikon N-90s, and the Canon EOS-1N and Canon 400mm lens with 1.4 extender. Nikon lenses used were: 17-35mm, 28-70mm, 80-200mm, and 600mm along with 1.4 extender. About 90 percent of the book photos were taken on 35mm film with the rest shot on Hasseblad gear. • Other pieces of equipment used were Nikon SB-28 strobe, Gitzo tripods, and Slik Head. Filters used were 81A and 85B. • Protect your lens from steam in thermal areas by using filters or covering the lens with a piece of clothing. I used Galen Rowell split neutral density filters. A special word of thanks to Galen Rowell. Sadly, he and his wife died in a plane crash. In my opinion his work is right at the top for adventure photography. I highly recommend his books, especially "Galen Rowell's Inner Game of Outdoor Photography." • All pictures were on slide film, 35 and 120mm film used were Kodak VS and SW both 100 ISO speed. Fuji films used were Provia 100 ISO speed and Velvia 50 ISO speed. • While film and gear help the picture taking process, in the end it's the photographer's way of seeing or vision that makes great photographs. In Yellowstone National Park there are many wonderful places to photograph. Weather changes rapidly and animals while more likely to be seen near sunrise or sunset, can appear at any time. No double exposures were done in camera or on a computer. • To buy pictures for either personal or commercial use or to contact me, see below.

Todd Anderson

2719 Oxford St.

Orlando, FL 32803

www.andersonpix.com

email: toddpix@aol.com

Acknowledgments

I am grateful for the help of many people who assisted in the publishing process: Bill Strode of Harmony House Publishing for outstanding design and editing and Shirley Williams for copy editing. Karen Boone and Joe Paul Pruett for design/editing. I appreciate the able counsel of good friend Ben Van Hook who helped picture edit from thousands of slides. Good friends, Mark Ashman, Jolie Johnson, and Sheri Lowen were most kind for driving me to and from the airport or giving me advice on this book. I met many wonderful employees and visitors in Yellowstone. The rangers who worked very hard to balance conflicting demands made my work easier. Distinguished photographer, Henry (Chip) Holdsworth, and friend, Tom Adams, aided with the best locations for grizzly bears. I must thank Frances Pokorny for her love of Yellowstone's special places and her affirmation of the park's unique place in the world. There are many other talented photographers and editors over the years who have helped me with the challenging profession of photography. David Roark with Walt Disney World Photography gave me many assignments. Finally I must thank my parents, Neil and Sarilda Anderson, and my sister, Ann Anderson. To put it simply, family is so important in my life. May Yellowstone always be protected for families and individuals to enjoy it.

Following page:
Shoshone Lake makes
a beautiful backcountry
camping spot.